Chemistry Experiments

Grades K–2

By
Wendi Silvano

Illustrations by
Gary Mohrman

Published by Instructional Fair
an imprint of

McGraw Hill **Children's Publishing**

M000024361

Author: Wendi Silvano

Children's Publishing

Published by Instructional Fair
An imprint of McGraw-Hill Children's Publishing
Copyright © 2004 McGraw-Hill Children's Publishing

Send all inquiries to:
McGraw-Hill Children's Publishing
3195 Wilson Drive NW
Grand Rapids, Michigan 49544

Hands-On Chemistry Experiments—Grades K–2
ISBN: 0-7424-2746-3

1 2 3 4 5 6 7 8 9 MAL 09 08 07 06 05 04
The McGraw-Hill Companies

Hands-On Table of Contents

0-7424-2746-3 *Hands-On Chemistry Experiments*

Hands-On Introduction to This Book

This book is designed to help your students build on their natural curiosity about the world around them. They will learn how to use the tools and methods of science to look for answers to their questions. The investigations described here support the new National Science Education Standards in their emphasis on science as a *process of inquiry*, rather than a recitation of facts.

Each activity includes an instruction page that is followed by a student reproducible page. Instruction pages begin with a question or guiding statement. This is followed by a materials list, step-by-step directions, suggestions for discussion, and helpful tips. The student pages feature a variety of activities that can be customized to the ability level of your students

This book is based on the National Science Education Standards for Physical Sciences. Chemistry is an integral part of the Physical Sciences. It is the study of what things are made of, how they interact with each other, and how they change. However, chemistry is not confined to the Physical Sciences. Chemical processes are also important in the Life Sciences and the Earth and Space Sciences. Some of the chemistry activities and the extensions found at the end of each section link to these other NSE standards.

This book has been divided into four sections: Matter, Mixtures and Solutions, Physical Changes and Chemical Changes. At the end of each section, you will find a page of suggestions for further inquiry in that field. Feel free to mix activities from different sections and adapt or customize the experiments and the student pages to your particular class.

0-7424-2746-3 *Hands-On Chemistry Experiments*

Hands-On Matter

In this section, students will have the opportunity to investigate the properties of matter. Matter is anything that takes up space and has weight. The study of chemistry is the study of matter since all substances and objects in our world are composed of matter.

Through the activities in this section, students will have the chance to identify matter through their different senses, investigate the question of whether air takes up space, where air is found, and how it is used. They will explore for themselves how liquids (primarily water) differ from solids, how water molecules attract one another and how that helps different materials to absorb at varying rates.

Students will use a variety of tools and methods to facilitate their explorations. They will record results, predict outcomes, develop their own questions to be answered, connect concepts with real-life situations, and even have a water drop race.

"Objects have many observable properties. Through the observation, manipulation, and classification of common objects, children reflect on the similarities and differences of the objects."

National Science Education Standards

Experiment 1 — Hands On

Objects have observable properties—identifying matter.
Can you identify substances by sight and touch?

What You Need:

- Six plastic or paper bowls for each group
- Paper towels
- Copy of page 7 for each student
- Marker
- Masking tape
- Pencils
- Six different substances (e.g. shampoo, honey, vegetable oil, butter, glue, batter, etc.)

What to Do:

1. For each group, prepare bowls by marking them #1–6 with masking tape.
2. Place one substance in each bowl.
3. Give each student the record sheet (page 7).
4. Have students look at and touch each substance, guess what it is, and record their guess on their record sheets.
5. When everyone has guessed, tell them what each thing is, and compare results.

Let's Talk About It:

What helped you to identify these bits of "matter"?

Teacher Tips:

Safety Tip: Make certain that students do not try to taste any of the substances.

The next activity (*Sniff It, Whiff It*, page 8) goes hand-in-hand with this activity.

Name _____

Date _____

Hands On Record Sheet

Draw or write what you think is in each bowl.

#1	#2
#3	**#4**
#5	**#6**

0-7424-2746-3 *Hands-On Chemistry Experiments*

Experiment 2 Sniff It, Whiff It

Objects have observable properties—identifying matter.
Can you identify matter by smelling it?

What You Need:

- Six different items that have strong smells (e.g. shampoo, onion, vanilla flavoring, peppermint flavoring, soy sauce, chocolate, piece of banana, perfume, etc.)

- Six opaque film canisters per group

- Copy of page 9 for each student

- Marker

- Pencils

- Awl or nail

What to Do:

1. For each group prepare film canisters by poking a small hole in the lid with an awl or nail (about the diameter of a pencil). Mark the canisters #1–6. Fill each with a different item to smell.

2. Give each student the record sheet (page 9).

3. Have students smell each container, guess what is inside and record their guess on their record sheets.

4. When everyone has guessed, tell them what each substance is and compare results.

Let's Talk About It:

What helped you identify these bits of matter?

Teacher Tips:

Safety Tip: Make certain that students do not try to taste any of the substances from the containers.

Name _____

Date _____

Sniff It, Whiff It Record Sheet

Draw or write what you think you smell.

#1	#2
#3	**#4**
#5	**#6**

0-7424-2746-3 *Hands-On Chemistry Experiments*

Experiment 3 Air Everywhere

**Substances have observable properties—air takes up space.
Is air real? Does it take up space?**

What You Need:

- Bread bags, produce bags, plastic grocery bags—one for each student

- Party horns (the type that honk)—one for each student if possible or several to share

- Copy of page 11 for each student

What to Do:

1. Ask: *What do you know about air? How do we know that air is real?*

2. Lay the bags out flat on the floor or table. Have students describe the shape of the bags (flat and thin). Ask: *Is there anything inside the bags?*

3. Give each student one of the bags and have them try to "catch" air in their bags by moving it through the air and twisting the top tightly to keep the air inside.

4. Have students feel the puffed-out bags. Ask: *What is inside? Does it take up space? How is the bag different now from before?*

5. Give each child a party horn. Have students see if they can blow the party horns with the air they have captured in their bags.

Let's Talk About It:

How could you tell that air takes up space?

Teacher Tips:

Hot dog/hamburger bun bags work well, as do plastic produce bags. Have a variety of bags available for experimentation and in case of holes.

Safety Tip: Do not allow children to place the bags over their heads or blow the party horns with their mouths.

0-7424-2746-3 *Hands-On Chemistry Experiments*

Name _____

Date _____

Where Is Air?

Where else could you catch air? Draw two different places in the boxes.

0-7424-2746-3 *Hands-On Chemistry Experiments*

Experiment 4 Bubble, Bubble

Substances have observable properties—air occupies space.
Does air take up space? How do we know?

What You Need:

- Bubble liquid—purchased or homemade.

- Pliable plastic straw for each student

- Paper cup for each student

- Pencil or crayons

- Scissors

- Copy of page 13 for each student

If you want to make your own bubble liquid, use this recipe: Mix 1 cup water, ¼ cup dishwashing liquid, ¼ cup glycerin, and a pinch of sugar with an eggbeater.

What to Do:

1. Put some bubble liquid into each student's cup.

2. Have students cut the pliable end of their straws into four parts and push back the parts (as in illustration). Bend the straw into a J-shape.

3. Have students dip the cut end into the bubble liquid and pull it out again.

4. Have them blow gently into the top of the straw, and experiment with making all kinds of bubbles.

Let's Talk About It:

What is inside the bubbles? How did it get in? Do bubbles help us see if air takes up space?

Teacher Tips:

Safety Tip: Do not allow students to drink or suck in the bubble liquid.

0-7424-2746-3 *Hands-On Chemistry Experiments*

Name _____

Date _____

Big Bubbles

What is the biggest bubble you can blow? Draw it here.

What did you have to do to blow big bubbles? _____

_____

Experiment 5 Is There Air in There?

**Objects may consist of one or more materials—some solid objects contain air.
Is there air in some solid objects?**

What You Need:

- Plastic tubs or plastic shoe boxes—one for each group

- Water

- Several small potted plants in dry soil—two for each group

- Newspaper

- Copy of page 15 for each student

- Pencils

- A variety of objects to test (e.g. sponges, sticks, rocks, brick, cardboard, pinecones, leaves, cloth, chalk, paper, crayons, etc.)

What to Do:

1. Fill the tubs or shoe boxes ¾ full of water.

2. Have one student in each group slowly immerse the potted plant in water. Observe the air bubbles that appear. Ask: *Where did the bubbles come from? What is in the bubbles? Where did the air come from?*

3. Ask: *Do you think the pot, the plant, or the soil produced the bubbles?* (The soil should produce the most bubbles.)

4. Allow students to test a variety of other objects to see which ones contain the most air. Have them draw or write the object to be tested on the *Checking for Air* prediction page (page 15), make a prediction, test the object, and record the results.

Let's Talk About It:

Why do some objects produce more bubbles than others?
How does this test help us see if there is air inside an object?

0-7424-2746-3 *Hands-On Chemistry Experiments*

Name _____

Date _____

Checking for Air

Check more things for air. Draw or write what you will check. Circle if you think there will be **lots** of bubbles, a **few** bubbles, or **no** bubbles. Then check for air. Place a check mark under the actual result.

What I Will Check	What I Think Will Happen			What I Think Will Happen		
	lots	few	none	lots	few	none
			X			

 0-7424-2746-3 *Hands-On Chemistry Experiments*

Experiment 6) Air It Out

**Substances have observable properties—air can be used to dry things.
How can air be used? Where will things dry more quickly?**

What You Need:

- Cloth squares, washcloths, or washable doll clothes for each group

- Plastic tubs with water—one for each group

- Copy of page 17 for each student

- Resealable bags, plates, clothespins, string, newspapers, or whatever else students need to carry out their tests

What to Do:

1. Ask: *Can you think of ways air helps us?*

2. Tell the students: *Today you will explore how air helps to dry things.*

3. Ask them to devise a plan to test how quickly things will dry in different conditions. Discuss possible ideas such as inside or outside, in the sun or the shade, hanging or laying down, in a plastic bag, under a plate, in between two papers, on top of various surfaces, etc.

4. Provide each group with a tub of water and several cloth squares.

5. Have students soak the cloths and test the drying rates.

6. Have students share their results with the class (pictures, oral report, chart, etc.).

Let's Talk About It:

What did you discover about how air helps to dry things?

Name _____

Date _____

What Dries Faster?

Cut out the pictures. What will dry first? What will dry last? Glue them onto a paper in order.

0-7424-2746-3 *Hands-On Chemistry Experiments*

Experiment 7 — Let It Flow

Substances have observable properties—water flows to fill the shape of the container.
How does liquid water compare to solid water (ice)?

What You Need:

- A variety of containers in different shapes and sizes (jars, cups, pans, vases . . .)
- Pitchers—one per group
- Bowls—one per group
- Water
- Ice cubes
- Copy of page 19 for each student

What to Do:

1. Ask students what they know about water and ice. *What is ice? How are water and ice different?*
2. Give each group a pitcher of water, a bowl of ice, and an assortment of containers.
3. Have each group experiment with pouring the water and the ice from one container to another.
4. Remind the students not to pour the ice and the water into the same container at this point.

Let's Talk About It:

A *liquid* is a substance that flows freely. A *solid* is something that keeps its shape. *Flows* means "to move freely." Listen to see if the students use these terms. If they describe these concepts in other ways, you can introduce the terms to them.

Name _____

Date _____

Fill 'Em Up

If you put the water from the glass into each thing, how high do you think it will go? Color it in.

0-7424-2746-3 *Hands-On Chemistry Experiments*

Experiment 8 Water Loves Water

Substances have observable properties—water molecules are attracted to each other.
What happens when water gets near other water?

What You Need:

- One sheet of waxed paper for each student—about 10 inches long

- Toothpicks—one per student

- Eyedropper—one per student

- Water in paper cups

- Straws—one per student for droplet race (page 21)

- Copy of page 21 for each student (or for each pair of students)

What to Do:

1. Give each student the waxed paper and a toothpick.
2. Have students place four or five small drops of water onto the paper (not too close together).
3. Students then dip the tips of their toothpicks in water.
4. Instruct them to bring the toothpicks very close to one of the drops without touching it.
5. Ask: *What happens? Is it the same with all the drops? Can you drag the drops across the paper? Can you drag all the drops together?*
6. Allow students to experiment with the water and the toothpicks.

Let's Talk About It:

What happens when two drops are very close to each other?
The drops of water will move toward the toothpick and "jump" together because water molecules have an attraction for each other. This is called *cohesion.*

Name _____

Date _____

Ready, Set . . . Go!

Lay this racetrack on your desk or a table. Lay a sheet of waxed paper on top of the track. Drop one drop of water on each side at the START. Use your straws to blow your drops to the FINISH. Get a friend to race with you. Race as many times as you want.

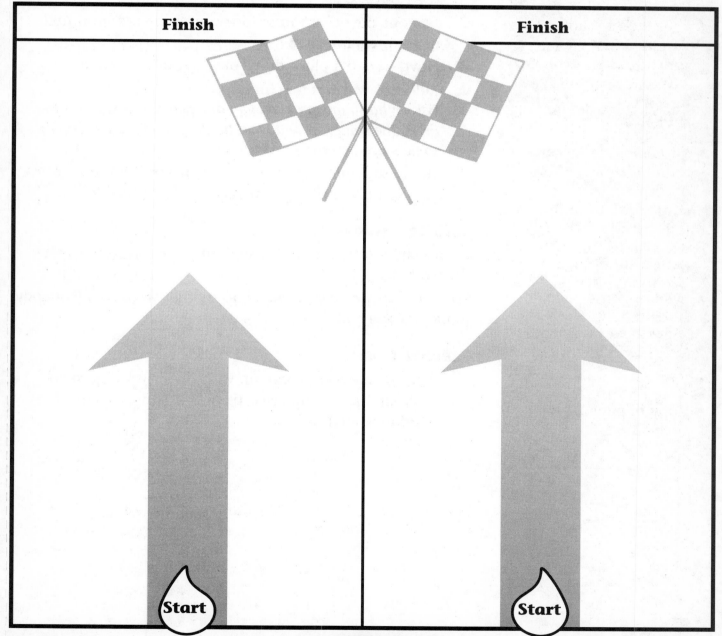

0-7424-2746-3 *Hands-On Chemistry Experiments*

Experiment 9 — Soak It Up

Substances have observable properties—capillary action of water.
What happens when water drops on different materials?

What You Need:

- Four-inch squares of a variety of papers (copy, construction, newsprint, waxed, paper towel)—one of each kind per student

- Various pieces of other materials to check for absorption (tissues, cardboard, aluminum foil, sponges, plastic, washcloths, cotton balls, rubber, etc.)

- Eyedroppers—one per student

- Paper cups

- Water

- Copy of page 23 for each student

What to Do:

1. Give each student a cup of water, an eyedropper, and their papers.

2. Lay out the other materials around the room on tables or counters for later.

3. Have students place drops of water onto each kind of paper.

4. Ask: *What happens when you dropped the drops? Are the drops the same? Why do you think some disappeared and some stayed on the surface?*

5. Have students test the materials around the room and record their results on the record sheet (page 23).

Let's Talk About It:

Discuss how some materials soak up or *absorb* water better than others.

What uses do we have for materials that absorb well? What about those that don't absorb?

Teacher Tips:

Allow students to come up with their own objects to test (e.g. their skin, desktops, shoes, sand, or soil, if available, etc.).

Name _____

Date _____

Does It Soak Up?

What else will soak up water? Try some things and see. Draw or write what you tried in the right boxes.

Soaks Up Well

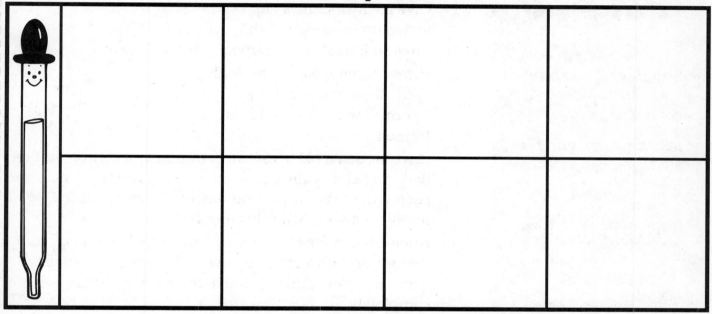

Does Not Soak Up Well

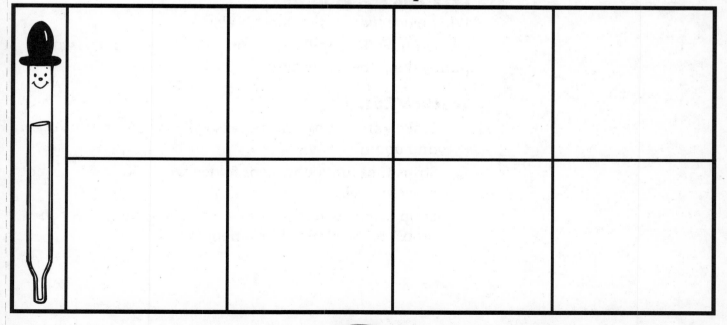

0-7424-2746-3 *Hands-On Chemistry Experiments*

Experiment 10 Wiggling Worms

Substances have observable properties—capillary action of water.

Can water make paper move?

What You Need:

- Construction paper cut into thin strips about ¼–½-inch wide and six inches long

- Eyedroppers—one per student

- Water in paper cups

- Copy of page 25 for each student

What to Do:

1. Give each student a cup of water, an eyedropper and three to four strips of paper.

2. Have students accordion fold their strips of paper into "worms" with folds of about ½ inch.

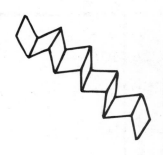

3. Have students place one strip of paper at a time on their desk and slowly drop just one or two drops of water onto each end of the worm and one or two drops in the middle. Observe what happens.

4. After a few moments they can slowly release drops along the rest of the worm. See how long it takes for enough water to be absorbed to make the worms flatten completely out.

Let's Talk About It:

What made the "worms" move? Why?

Discuss how the capillary action (soaking up) of the water pulled the paper folds apart.

Teacher Tips:

Make extra strips, as students will want to try this again and again.

Strips that are too wide or folded with wide folds do not move as well.

Drop on the water a little at a time (waiting between drops) to maximize movement.

Soaking Wet

Sometimes things that soak up water will help us. Sometimes we want something that does not soak up water. Look at the pictures on the left. Draw lines to match them to something on the right that would help.

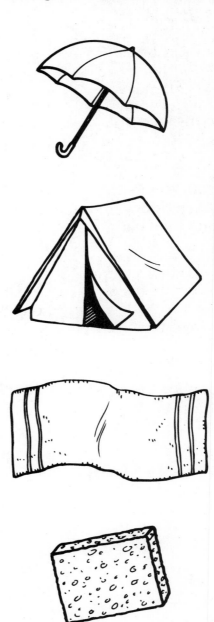

0-7424-2746-3 *Hands-On Chemistry Experiments*

Changing Colors

Substances have observable properties—capillary action in water.
Organisms have different structures that serve different functions in growth.
How does a plant soak up water?

What You Need:

- Water

- Red food coloring

- Three clear jars or glasses per group

- 2" x 6" strips of paper towel, light-colored flower such as a carnation (one per group)

- Stalk of celery—one per group

- Transparent tape

- Pencil

- Copy of page 27 for each student

What to Do:

1. Put one to two inches of water into each jar.

2. Put two drops of red food coloring into two of the jars.

3. Cut the stems of the flower and the celery and stick each into a jar with coloring.

4. Put one drop of food coloring near one end of the paper towel.

5. Tape the other end of the paper towel to the center of the pencil and roll it around the pencil a few times. Place the pencil across the top of the jar, allowing the colored end of the paper towel to hang just in the water.

6. Observe and watch the capillary action as the color travels upward.

Let's Talk About It:

Capillary action is the science name for "soaking up." *How does the capillary action of the water on the paper towel compare to that in the plants?*

Teacher Tip:

For a technology element to this experiment, allow the children to come up with their own way of hanging the paper towel in the water.

0-7424-2746-3 *Hands-On Chemistry Experiments*

Name _____

Date _____

Movie Magic

Your teacher will tell you what to do to make a little movie book of what happened with the water.

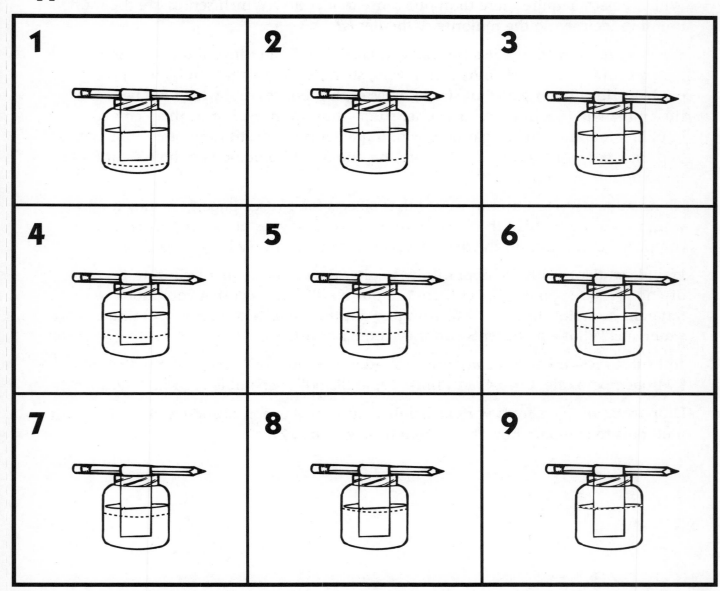

Instructions to be read aloud to students: First, color the pictures of paper towels in each square up to the dotted line with a red crayon. Then cut the pages apart, put them in order, and staple them on the left side to make a mini-book. To watch your "movie," just flip the right edge of your book quickly to thumb through the book in one fast motion.

0-7424-2746-3 *Hands-On Chemistry Experiments*

Have your students bring in "mystery matter" from home (an object or substance they bring in a film canister, or other opaque container—see pages 6 and 8). These can be used in an additional class activity or in a center where the children try to identify the matter using just sound, touch, smell, etc., or a combination of senses. Which objects require more than one sense to identify? Which senses are the most useful in identifying the majority of the items?

When working with bubbles (see page 12), ask children to invent different tools (using common items such as string, wire, sticks, etc.) to help with blowing the bubbles. Have them see what size bubbles each produces and which are best. Can they make bubbles that are any other shape than round? Change the recipe for bubble solution by altering amounts of ingredients or substituting other ingredients and then compare qualities. Compare the quality of homemade bubble solution to those bought in stores.

Allow students to test a number of other objects for air (see page 14). They might want to test a variety of the same item (e.g. various types of soil or paper) and compare amounts of air in each. How could you test a very large object for air?

Provide students with the opportunity to alter the variables involved in testing the drying rates (see page 16). Let them test during different weather conditions, with different materials to soak, with wringing out the water first vs. not wringing out the water, etc. What experiments can they come up with?

Test substances other than water for cohesion (see page 20). Try rubbing alcohol, hydrogen peroxide, Coca-Cola®, milk, Pepto Bismol®, vegetable oil, Karo® syrup, etc.

Drop water droplets from various heights onto various non-absorbent and absorbent materials to compare how the drops look (see page 22).

In this section, students will investigate mixtures and solutions. A *mixture* is a combination of any two or more materials. In a mixture the pieces are combined, but each substance keeps its original chemical properties. A *solution* is a special kind of mixture in which one substance (the solute) is dissolved in another substance (the solvent). In a solution, the molecules are mixed up in a completely even distribution.

Through the activities in this section, students will have the opportunity to prove that water molecules are constantly moving, will make and explore both mixtures and solutions, test the solubility of several substances, and even create fascinating crystal sticks by removing a solute from a solution.

Your students will use Earth materials such as soil and water to explore solubility. They will see how such familiar items as the food they eat are mixtures, use a ruler and their artistic creativity to make straight line designs that resemble crystal sticks, and even practice their language arts skills as they use verbs to describe the movement they see while dissolving a powdered fruit drink.

"In the earliest years, investigations are largely based on systematic observations."

National Science Education Standards

Experiment 12 Moving Molecules

Substances have observable properties—water molecules are always in motion.
Do water molecules move all the time?

What You Need:

- A clear glass or jar
- Water
- Red food coloring
- Copy of page 31 for each student

What to Do:

1. Fill the glass or jar about ¾ full of water. Allow it to sit undisturbed all day.

2. On the second day, slowly add three or four drops of red food coloring. Do not bump or move the water in any way. Observe the drops as they fall through the water.

3. Allow the glass or jar to remain undisturbed the rest of the day.

4. On the third day, check the water.

Let's Talk About It:

Why do you think on the third day the coloring has spread evenly throughout the water? (The constant movement of the water molecules pushes the food coloring particles around until they are spread throughout the water. This process is called *diffusion*.)

What could you do to diffuse the color more quickly? (Stir the water)

Moving Molecules Report

Draw or write what happened in each step.

How did the water look before we put in the color?

How did the water look when we first put in the color?

How did the water look after it sat for one day?

Experiment 13 Mix It Up

Substances interact with other substances—making a mixture.
What is a mixture?

What You Need:

For each student or group

- One cup of water
- One cup of flour
- One cup of salt
- Food coloring
- Spoon
- Bowl
- Large resealable plastic bag
- Measuring cup
- Copy of page 33 for each student

What to Do:

1. Tell students that they will be making a mixture of three substances (water, flour, and salt).
2. Put one cup of flour and one cup of salt into the bowl. Mix.
3. Add a drop or two of food coloring to the water.
4. Gradually add the colored water, a little at a time, to the bowl. Stir. Add more water until the clay is completely mixed but not too sticky. If it is too sticky, add a little more flour.
5. After experimenting with the play clay, store it in a resealable plastic bag.

Let's Talk About It:

Is there still flour, salt, and water in our play clay? (Yes)

The flour, salt, and water are mixed together, but each one still has the same chemical properties as before. This is called a *mixture*.

Teacher Tips:

Save the play clay for use in "Playing with Play Clay."

Mixtures on the Menu

Many of the foods we eat are mixtures. Do you know some of the parts that make up these foods? Write or draw them in the boxes.

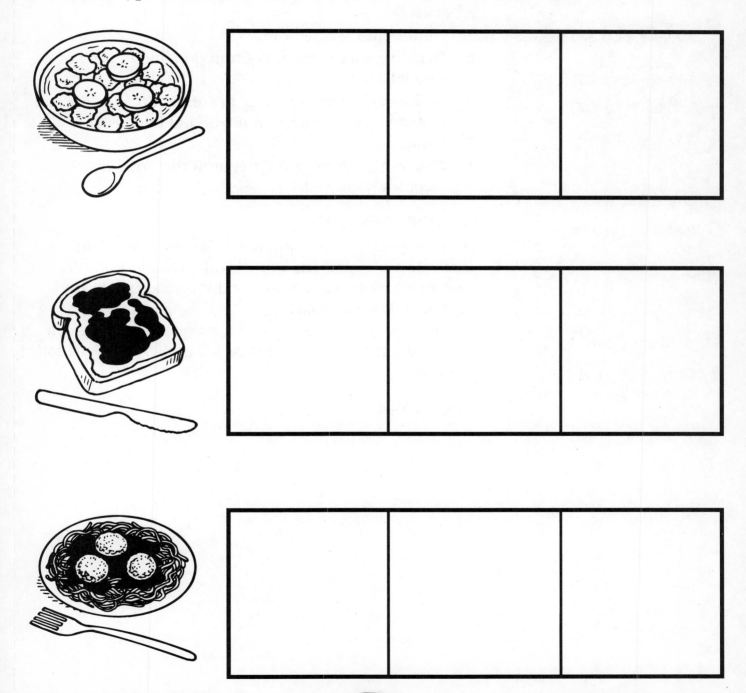

0-7424-2746-3 *Hands-On Chemistry Experiments*

Experiment 14 Swirl and Twirl

**Substances interact with other substances—making a solution.
Can you observe how one substance dissolves in another?**

What You Need:

- Transparent glasses—one per group or student

- Water

- Craft sticks—one per group or student

- Red powdered fruit drink

- Copy of page 35 for each student

What to Do:

1. Fill the cups ¾ full of water.
2. Pick up a small amount of fruit drink on the end of the craft sticks.
3. Very gently shake a little of the drink mix over the cup of water and observe as it mixes and then dissolves in the water.
4. Add more, a little at a time, until the water is completely colored.

Let's Talk About It:

The powdered drink crystals are called the *solute*. They *dissolve* in the water. The water is called the *solvent*. When a substance dissolves in a solute and is spread evenly, the result is called a *solution*.

0-7424-2746-3 *Hands-On Chemistry Experiments*

 **Mixtures and Solutions**

Name _____

Date _____

Swirl and Twirl

Draw how the powder moved in the glass.

Write three words that tell how the powder moved in the water.
(Example: *Twirl*)

_____ _____ _____

0-7424-2746-3 *Hands-On Chemistry Experiments*

Experiment 15 · Dissolve or Not Dissolve?

Substances interact with other substances—solubility of substances.
What will dissolve in water and what will not?

What You Need:

- Water

- Plastic cups—three for each group

- Salt

- Pepper

- Soil

- Spoons—three for each group

- Copy of page 37 for each student

What to Do:

1. Fill the cups about ¾ full of water.

2. Put one spoonful of salt into one cup. Put one spoonful of pepper into another. Put a spoonful of soil into the third.

3. Stir each one for a few minutes to observe if they dissolve in water.

Let's Talk About It:

Which one dissolved? (Salt). *What happened to the soil?* (It sank to the bottom.) *What happened to the pepper?* (It floats.) When a substance floats in a liquid, but is not dissolved, this is called a *suspension*.

Name _____

Date _____

Did It Dissolve?

Draw what it looked like in each cup when you put something into the water.

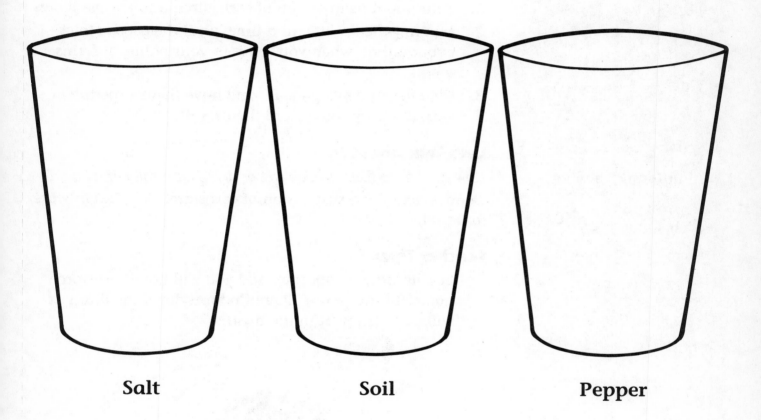

Salt   Soil   Pepper

Bonus Questions

Which one dissolved (is a solution)? _____

Which one floated (is a suspension)? _____

Which one sank (is only a mixture)? _____

0-7424-2746-3 *Hands-On Chemistry Experiments*

Experiment 16 Goop

Substances interact with other substances—exploring a unique mixture.
What happens when cornstarch is mixed with water?

What You Need:

- Cornstarch
- Water
- Paper towels
- Spoons
- Aluminum pie plates
- Copy of page 39 for each student

What to Do:

1. Pour about ½ to ¾ cup of cornstarch into the pie plates.
2. Mix in water, a little at a time, until the *goop* is thick enough that, when you roll it in your palms, it forms a ball.
3. Give the students page 39 and have them experiment with the goop and record their results.

Let's Talk About It:

How does goop feel? Is goop wet or dry? Is it a solid or a liquid?
Goop is actually a suspension of cornstarch (a solid) in water (a liquid).

Teacher Tips:

Let the water evaporate, and you will have cornstarch again for future use. Do not pour it down the drain. It will clog! Throw it in the trash.

Name _____

Date _____

Goop Tests

Try these things with *goop*. Does the goop act more like water or clay each time? Circle what you think. Try your own ideas. Write what you did.

Roll it	water	clay
Slap it	water	clay
Squeeze it	water	clay
Put your finger in slowly	water	clay
Set a pencil on top of it	water	clay
_____	water	clay
_____	water	clay
_____	water	clay
_____	water	clay

0-7424-2746-3 *Hands-On Chemistry Experiments*

Experiment 17 — Crystal Sticks

Substances interact with other substances—removing a solute from a solution.
How can you remove a solute from a solution?

What You Need:

- Epsom salt
- Black construction paper
- Water
- Margarine tub lid
- Spoon
- Scissors
- Jar or glass
- Tape
- Magnifying glass, if available
- Rulers and crayons
- Copy of page 41 for each student

What to Do:

1. Fill the jar or glass half full of water.
2. Add two to three spoonfuls of epsom salt to the water. Stir until mostly dissolved.
3. Cut a circle of black paper to fit inside the margarine tub lid. Tape it inside the lid.
4. Slowly pour a thin layer of the epsom solution over the paper.
5. Set it where it will not be disturbed for several days.
6. Observe the crystals that remain after the water evaporates, if possible with a magnifying glass.

Let's Talk About It:

How did we get the solute (epsom salt) *out of the solution?* (Allow water to evaporate.) *What is left?* (epsom salt crystals).

0-7424-2746-3 *Hands-On Chemistry Experiments*

Straight Line Art

The crystals we made are all thin, straight lines. Use a ruler and some crayons and see what kind of interesting picture you can make here using only straight lines.

© McGraw-Hill Children's Publishing

0-7424-2746-3 *Hands-On Chemistry Experiments*

Try mixing several drops of different food colorings in small containers of water and use them to paint with like watercolors. Which colors are the most brilliant? How many drops of coloring does it take to make a color bright enough to see?

What other homemade dough/mixture recipes can you make? How do their characteristics vary? Allow students to develop and compare their own recipes using such substances and flour, salt, alum, cornstarch, wheat paste, liquid starch, vegetable oil, cream of tartar, baking soda, etc.

After dissolving the powdered drink mix in water, try it again, adding the sugar to make the drink. Make other homemade drink solutions such as homemade lemonade, chocolate milk, and fruit smoothies. Try the same experiment using different colors of fruit drink powder, mixing colors together (e.g. red and yellow to get orange, blue, and yellow to get green, etc.).

After completing the solubility test on salt, pepper, and soil, try other materials to see what will dissolve in water and what will not? Is there anything you can do that will get something to dissolve that did not dissolve in your test?

After experimenting with goop, try out various proportions of water and cornstarch to see what differences occur in the result. Are there other substances that react similarly to goop when mixed with water? What else could you try?

After making the crystal sticks from epsom salts, use other solutes (salt, sugar, rock salt, borax, baking soda, alum, etc.) to see if they make crystals when removed from their solutions.

Hands-On Physical Changes

This section deals with physical changes. Physical changes occur when a substance changes from one form to another or one state to another. However, the chemical make-up of the substance itself does not change. Physical changes can occur in many ways (cutting, bending, melting, freezing, adding water, etc.).

Through the activities in this section, students will explore and create a variety of physical changes in several different substances. They will decide on their own ways to physically change paper and play clay, observe a physical change that occurs as they make pudding, and compare how changing sponges by adding water changes their weights. They will experience the three main states of matter as they melt and refreeze Popsicles™, inflate a balloon with dry ice as it changes states, discover that gas molecules can seep through balloons, and even enjoy the physical change that occurs when they pop popcorn.

"During the early years, children explore the world by observing and manipulating common objects and materials in their environment. Children compare, describe, and sort as they begin to form explanations of the world. Developing a subject-matter knowledge base to explain and predict the world requires many experiences over a long period."

National Science Education Standards

Experiment 18 Bend It, Change It

Materials have observable properties—observe the effects of physical changes.

How many ways can you change a piece of paper?

What You Need:

- Several sheets of paper for each student
- Scissors
- Water
- Pencil
- Copy of page 45 for each student

What to Do:

1. Give each student several sheets of paper.

2. Ask: *What are some ways you can change a piece of paper?* Possibilities might include cutting, bending or folding, tearing it in bits, crinkling, writing on it, soaking it, etc.

3. Allow students to try a variety of things and record their results on the record sheet.

Let's Talk About It:

After each change, was the paper still paper? How was it different? How was it the same?

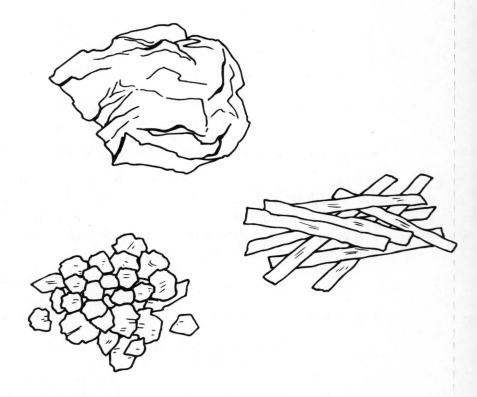

Name _____

Date _____

Bend It, Change It Record Sheet

In each box, draw what the paper was like after you changed it. Write what you did to the paper on the line.

_____	_____	_____
_____	_____	_____
_____	_____	_____

0-7424-2746-3 *Hands-On Chemistry Experiments*

Experiment 19 Playing with Play Clay

Materials have observable properties—observing physical changes.
What ways can you change play clay?

What You Need:

- Play clay saved from *Mix It Up* (page 32) or made from recipe on that page

- Utensils such as cookie cutters, plastic knives, and rolling pins for use in altering the play clay.

- Copy of page 47 for each student

What to Do:

1. Give students their bags of play clay.
2. Ask: *How can you change the play clay?* (Possibilities might include rolling, squashing, squeezing, poking with various objects, twisting, cutting, molding, crumbling, wetting, etc.)
3. Allow students to try different possibilities and record their results on the record sheet.

Let's Talk About It:

After you changed the play clay, was it still clay? How was it different? How was it the same?

Playing with Play Clay Record Sheet

In each box, draw what the play clay was like after you changed it. Write what you did to it on the line.

_____	_____	_____
_____	_____	_____
_____	_____	_____

0-7424-2746-3 *Hands-On Chemistry Experiments*

Experiment 20 — Shake It Up

Observing a physical change caused by shaking.
What change occurs when you shake pudding mix and milk?

What You Need:

- One box instant pudding for every four students

- Milk—½ cup per student

- One empty yogurt container with lid per student

- One plastic spoon per student

- Copy of page 49 for each student

- Measuring cup

What to Do:

1. Pass out yogurt containers and spoons.
2. Pour ¼ box pudding powder into each student's container. Add ½ cup milk.
3. Ask: *How would you describe this mixture?* (runny, liquid, etc.)
4. Have students put on the lids and shake vigorously for one minute.
5. Allow the pudding to set for three minutes.
6. Ask: *Now how would you describe the mixture?* (thicker, more solid)
7. Eat and enjoy!

Let's Talk About It:

What kind of change occurred when we shook the pudding? (Physical change). *Is it still made of milk and pudding powder? How is it different?*

 0-7424-2746-3 *Hands-On Chemistry Experiments*

Name _____

Date _____

Making Pudding

Cut out the strips that tell what we did to make pudding. Paste them in order inside the squares or draw what we did in each step.

1	**2**
3	**4**

We put pudding mix and milk in our cups.

We ate the pudding.

We let it set for three minutes.

We shook our cups.

0-7424-2746-3 *Hands-On Chemistry Experiments*

Experiment 21 — The Weight of Water

Objects can be measured—comparing weights after a physical change.
What change can you make to a sponge that will make it weigh more?

What You Need:

- Sponges—two per group
- Balance scale—one per group
- Water
- Sink or bowl for water
- Copy of page 51 for each student

What to Do:

1. Give each group two sponges and a scale.
2. Place one sponge on each side of the scale. Ask: *Do they weigh the same? What change could we make to one of the sponges that would make it weigh more than the other one?* (Add water.)
3. Soak one sponge in water and place it on the scale.
4. Compare weights.

Let's Talk About It:

What change did we make to the sponge? (Added water.) *Is it still a sponge?*

Teacher Tips:

You might want to have the students try comparing the weight of a wet sponge that has been wrung out with one that has not.

Name _____

Date _____

Heavy or Light?

Color the one that would weigh more in each picture.

Experiment 22 Popsicle™ Puddles

Materials can exist in different states—observing the change from solid to liquid. What happens to a Popsicle™ when it is left out of the freezer for awhile and then returned to the freezer?

What You Need:

- Popsicles™
- Paper cups
- Freezer
- Markers
- Copy of page 53 for each student

What to Do:

1. Hold up one Popsicle™. Ask: *How would you describe this Popsicle™?* (hard, solid, frozen, etc.)

2. Have students write their names on paper cups. Give each student a Popsicle™ and have them put it in their cup upside down (stick pointing up).

3. Allow the Popsicles™ to melt. Ask: *Now how would you describe the Popsicle™?* (liquid, runny, a puddle, wet, etc.)

4. Discuss how the Popsicle™ has changed from a solid to a liquid.

5. Ask: *What could we do to make the Popsicle™ solid again?* (Freeze it.)

6. Put the cups with the melted Popsicle™ back into the freezer. Eat when frozen.

Let's Talk About It:

What two states of matter did the Popsicle™ go through? (Solid and liquid.) *Was it still the same substance?*

Teacher Tips:

To refreeze the Popsicle™ with sticks, you can cut the sticks in half and stick them back in when the Popsicle™ are about half frozen. Then peel off the paper cups when they are done.

0-7424-2746-3 *Hands-On Chemistry Experiments*

Name _____

Date _____

Melting and Freezing

Draw what the Popsicle™ was like in each step. Circle if it was solid or liquid.

Solid

Liquid

Solid

Liquid

Solid

Liquid

0-7424-2746-3 *Hands-On Chemistry Experiments*

Experiment 23 Solid to Gas

Materials can exist in different states—observing the change from solid to gas.
What happens to dry ice when it is left out of the freezer?

What You Need:

- One large balloon for each group

- One small chunk of dry ice per group

- Copy of page 55 for each student

- Glove for teacher to use when handling dry ice

What to Do:

1. Give each group a balloon.
2. Use a glove to place a small chunk of dry ice into each balloon.
3. Have students tie the ends of the balloons tightly.
4. Observe as the balloons begin to inflate.

Let's Talk About It:

Dry ice is a solid (frozen carbon dioxide). Outside the freezer, it will go from a solid state to a gaseous state directly without becoming a liquid. This process is called *sublimation*. The gas cannot escape the balloon, so it stretches it (inflating it).

Teacher Tips:

The chunk of dry ice should be about the size of a rubber pencil eraser. If it is too big, it will pop the balloon.

Safety Note: Always use gloves with dry ice. It will cause severe burns. Do not allow students to touch the dry ice.

Solid to Gas

Draw what the balloon looked like before and after.

Before

After

Experiment 24 — What's That Smell?

Materials exist in different states—detecting gas molecules escaping from a solid. Can gas molecules seep through balloons?

What You Need:

- Four balloons
- Funnel
- Extract of lemon, peppermint, vanilla, root beer concentrate
- Four small paper cups
- Copy of page 57 for each student
- Marker

What to Do:

1. Pour a little of each extract into a paper cup.
2. Allow students to smell each extract in the bottle. Ask: *Are the extracts solid or liquid?*
3. Mark the balloons #1–4.
4. Have students smell the balloons.
5. Ask: *Are the balloons solid or liquid?*
6. Using the funnel, place a little bit of each extract in a balloon.
7. Inflate and tie the balloons.
8. Wait 15 minutes and smell the balloons. Ask: *Can you tell what is in each balloon?*

Let's Talk About It:

Is the liquid leaking from the balloons? (No.) *Why do you think we can smell the extracts?* (The liquid extracts are beginning to evaporate, and the gas molecules are seeping through the balloon.)

Name _____

Date _____

What's That Smell Matching Sheet

What can you smell in each balloon? Match the balloons to the right smells.

Lemon

Peppermint

Vanilla

Root Beer

0-7424-2746-3 *Hands-On Chemistry Experiments*

Experiment 25 Popcorn Popping

Observing the effects of a change of state from liquid to gas.
What happens when you heat up a popcorn kernel?

What You Need:

- Hot air corn popper
- Popcorn kernels
- Glue
- Copy of page 59 for each student

If you do not have a hot air corn popper, you can use microwave popcorn. However, you should get a few popcorn kernels for the students to examine before the popping.

What to Do:

1. Have students examine the kernels of unpopped popcorn. Discuss its characteristics.
2. Put the kernels into the popper and watch them pop.
3. Ask: *What did we do to the kernels?* (heat them). *How have they changed?* (big and puffy). *Is it still corn?*
4. Save a few kernels for use with page 59. Eat and enjoy the rest.

Let's Talk About It:

There is a tiny bit of water inside the corn kernels. When it is heated up, it turns into a gas. The expanding molecules exert such pressure that it explodes, pushing out the puffy starch inside.

0-7424-2746-3 *Hands-On Chemistry Experiments*

Popcorn Popping Record Sheet

Glue or tape one unpopped and one popped corn kernel here. Then write to tell what each is like.

Before Popping

What It Is Like

_____

_____

_____

_____

_____

After Popping

What It Is Like

_____

_____

_____

_____

_____

0-7424-2746-3 *Hands-On Chemistry Experiments*

Have students watch for physical changes they make or see each day and make a class chart. Possibilities might include stretching their socks as they put them on, pouring milk on cereal, slicing bread or fruit, braiding hair, folding papers, making gravy, watering dirt, etc. Experiment with different materials and make physical changes of all sorts.

After comparing the weights of dry and wet sponges, have students compare weights of other materials before and after making various physical changes (e.g. a piece of chalk compared to the same-sized piece crumbled, a cup of juice compared to a frozen cup of juice, etc.).

After melting and refreezing the Popsicles™, make your own out of juice or powdered drink mixes. Use a variety of containers with different shapes and compare the frozen forms. Melt Popsicles™ or ice cubes in various places and situations to compare rates at which they melt (one in the sun, one in the closet, one in a shady spot, one by a fan, etc.).

Place any leftover dry ice from the activity *Solid to Gas* in a bowl with a small amount of water. Students can watch the escaping gases and feel how cool they are. You can also use the dry ice and some root beer concentrate from *What's That Smell?* to make homemade root beer. Just follow the instructions on the box of root beer concentrate.

After completing *What's That Smell?*, try various other substances that have odors. *Which ones are strongest? How many drops must you put in the balloon to still be able to smell it? Can you smell the extracts through plastic bags? Cardboard boxes? Plastic containers?*

Hands-On Chemical Changes

This section deals with chemical changes. Chemical changes occur when the chemical make-up of a substance completely changes. The atoms have been rearranged to create an entirely new substance with totally different properties than the original substances. Some form of energy (often heat) is needed to start a chemical reaction.

Through the activities in this section, students will experience a variety of chemical reactions with different materials. They will separate the solid and liquid parts of milk and then use the solids to make a homemade glue. They will be amazed and delighted to watch the top explode off a film canister, bubbling fizz pour out of a glass, and a balloon inflate by itself. They will test to see what vinegar does to coins and a raw egg. They will also use lemon juice to draw a hidden picture that will be revealed only through a chemical reaction.

"When carefully observed, described, and measured, the properties of objects, changes in properties over time, and the changes that occur when materials interact provide the necessary precursors to the later introduction of more abstract ideas in the upper grade levels."

National Science Education Standards

0-7424-2746-3 *Hands-On Chemistry Experiments*

Experiment 26 Lumpy Milk

Substances interact—separating milk into solid and liquid parts.
What happens when you mix milk with vinegar?

What You Need:

- Fresh milk
- Vinegar
- Spoon
- Glass or jar
- Copy of page 63 for each student

What to Do:

1. Fill the glass or jar with milk.
2. Add three spoonfuls of vinegar.
3. Let the mixture sit for several minutes.
4. Examine the results.

Let's Talk About It:

The solid particles in milk are not completely dissolved. They are spread evenly throughout the liquid. The vinegar causes them to clump together. These solid clumps are called *curd*. The liquid part is called *whey*.

Teacher Tips:

The curds (solid clumps) from this activity will be used in the next activity.

0-7424-2746-3 *Hands-On Chemistry Experiments*

Lumpy Milk Record Sheet

Draw what the milk was like before and after we put in the vinegar.

Before

After

0-7424-2746-3 *Hands-On Chemistry Experiments*

Experiment 27 Homemade Glue

Substances react with other substances—making homemade glue.
How can you make homemade glue?

What You Need:

- Curds from the previous activity
- Baking soda
- Strainer
- Bowl
- Scissors
- Regular school glue
- Copy of page 65 for each student
- Extra paper

What to Do:

1. Pour the milk from the *Lumpy Milk* activity into a strainer to separate the curds from the whey. Put the curds into a bowl and throw away the liquid.
2. Mix a teaspoon of baking soda with the curds. Stir until smooth.
3. Let the mixture sit overnight.
4. Try out your glue on page 65.
5. Test it against your regular school glue.

Let's Talk About It:

The curds (*casein*) from the milk react with the baking soda and produce a sticky substance.

 **Chemical Changes**

Name _____

Date _____

Glue Test

Cut out the shapes. Paste some on another paper with the homemade glue. Paste some with your regular glue. Which works best?

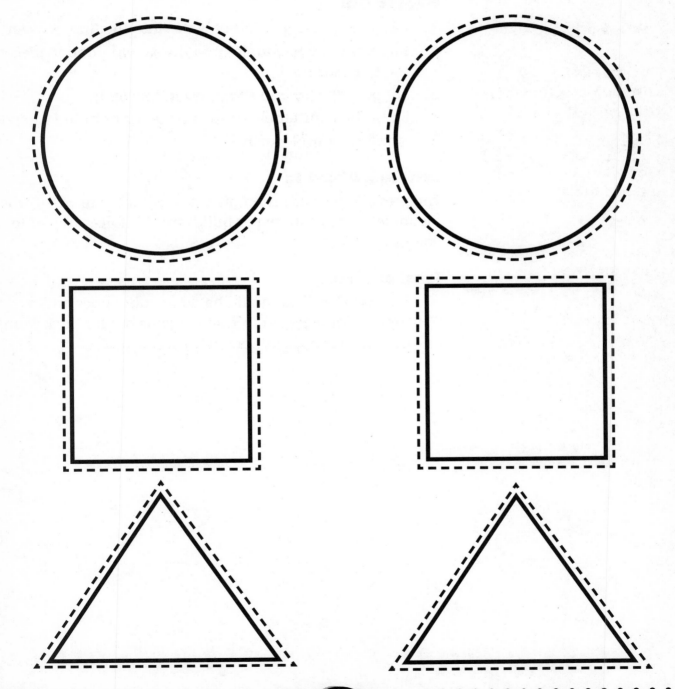

0-7424-2746-3 *Hands-On Chemistry Experiments*

Experiment 28 · Pop the Top

Substances interact with other substances—observing a chemical reaction.
What happens when a gas forms in a tightly closed container?

What You Need:

- 35-mm plastic film canister (preferably transparent)
- Eyedropper
- One effervescent antacid tablet
- Water
- Copy of page 67 for each student

What to Do:

1. Put one eyedropper full of water into the film canister.
2. Break the antacid tablet into thirds and put one piece in the film canister.
3. Snap on the lid and shake several seconds.
4. Place the canister on a flat surface far from the students.
5. Stand back and watch.

Let's Talk About It:

The pressure of the carbon dioxide gas that is released when the tablet mixes with water builds until it forces the lid to pop off.

Teacher Tips:

Safety Tip: This should be done only under adult supervision. Be certain that students are far back from the canister as it will pop with great force.

Name _____

Date _____

Pop the Top Record Sheet

Cut and paste to tell what happened.

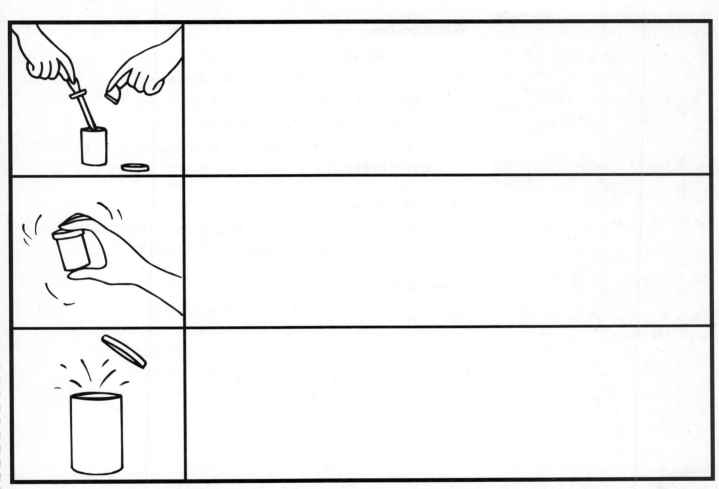

0-7424-2746-3 *Hands-On Chemistry Experiments*

We put the lid on and shook it up.

The lid popped off.

We put water and the tablet in the film case.

Experiment 29 Fizz

Substances interact with other substances—observing a chemical reaction.
What happens when you mix vinegar and baking soda?

What You Need:

- Transparent glass
- Shallow pan
- Vinegar
- Baking soda
- Food coloring
- Measuring cup
- Teaspoon
- Copy of page 69 for each student
- Scissors
- Glue
- Paper

What to Do:

1. Place the glass in a shallow pan.
2. Put in one teaspoon of baking soda.
3. Pour ½ cup vinegar into the measuring cup.
4. Add two drops of food coloring.
5. Dump the colored vinegar into the glass with the soda and observe.

Let's Talk About It:

The vinegar reacted with the baking soda and caused fizzy bubbles of carbon dioxide gas to be released.

Teacher Tips:

The next experiment *What Reacts?* is a good follow-up to this one.

© McGraw-Hill Children's Publishing

0-7424-2746-3 *Hands-On Chemistry Experiments*

First, Second, Third, Fourth

Cut out the squares and paste them in order to show what happened.

0-7424-2746-3 *Hands-On Chemistry Experiments*

Experiment 30 What Reacts?

Substances react with other substances—testing for reaction.
What liquids will react with baking soda?

What You Need:

- Four clear glasses for each group

- Baking soda

- Lemon juice, orange juice, milk, vegetable oil for each group

- Teaspoon for each group

- Copy of page 71 for each student

What to Do:

1. Ask: *What happened when we added vinegar to the baking soda in the last experiment?* (It reacted by making fizzing bubbles.)
2. Tell students that they will test some other liquids to see if they react, also.
3. Have students put one teaspoon of baking soda into each of the four glasses.
4. Add some of each liquid to test for a reaction (fizzing bubbles).
5. Record your results on the record sheet.

Let's Talk About It:

Liquids that contain acid will react. The reaction may not be as pronounced as with the vinegar, but there will be some fizzing or bubbles.

Teacher Tips:

This activity is much more effective after doing the activity *Fizz*. You may want to allow the students to choose their own liquids to test and make a class chart of the results. You could test as many liquids as they wish.

Name _____

Date _____

What Reacts with Baking Soda?

Circle "yes" or "no" to show if each liquid fizzed.

Lemon Juice

yes   no

Milk

yes   no

Vegetable Oil

yes   no

Orange Juice

yes   no

0-7424-2746-3 *Hands-On Chemistry Experiments*

Experiment 31 Blowing Up

Substances react with other substances—observing the reaction of yeast.
How does yeast react with water and sugar?

What You Need:

- Soda pop bottle or other bottle with skinny neck

- Warm water

- Dried yeast

- One teaspoon sugar

- Balloon

- Bowl

- Copy of page 73 for each student

- Measuring cup

What to Do:

1. Put the package of yeast in the bottle.
2. Add ¼ cup warm water.
3. Stir in the sugar.
4. Stretch the balloon over the neck of the bottle.
5. Fill the bowl with warm water and stand the bottle in it.
6. Observe for about 15–20 minutes.

Let's Talk About It:

Yeast is alive. It feeds on the sugar and produces carbon dioxide gas. This gas expands and begins to inflate the balloon.

0-7424-2746-3 *Hands-On Chemistry Experiments*

Blowing Up Record Sheet

This is what the balloon was like at first.	This is what the balloon was like after several minutes.

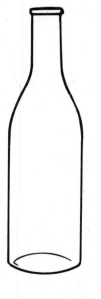

0-7424-2746-3 *Hands-On Chemistry Experiments*

Experiment 32 > Green Money

Substances interact with other substances—observing a chemical reaction.
How does vinegar react with copper?

What You Need:

- One penny, one nickel, one dime, one quarter
- Paper towel
- Paper plate
- Vinegar
- Copy of page 75 for each student

What to Do:

1. Fold the paper towel in half two times and put it on the paper plate.
2. Pour on some vinegar to soak the paper towel.
3. Place the coins on top of the paper towel.
4. Leave overnight and observe.

Let's Talk About It:

Vinegar is an acid (*acetic acid*) that combines with the copper in the coins and the oxygen in the air to create *copper acetate*. This is what coats them green.

0-7424-2746-3 *Hands-On Chemistry Experiments*

Name _____

Date _____

Colored Coins

Color in what happened to the coins. Write what happened.

_____

_____

_____

_____

_____

_____

_____

_____

0-7424-2746-3 *Hands-On Chemistry Experiments*

Experiment 33 Eggs-periment

Substances react with other substances—observing a chemical change.

What happens to a raw egg in vinegar?

What You Need:

- Glass jar with a lid
- Raw egg
- Vinegar
- Copy of page 77 for each student

What to Do:

1. Place the raw egg carefully in the jar.
2. Fill the jar with enough vinegar to cover the egg and close the lid.
3. Ask: *What do you see forming on the egg right now?* (bubbles)
4. Check the egg throughout the day and again after 24 hours.
5. Ask: *Can you see the yolk? Are there any pieces of shell floating in the vinegar?*

Let's Talk About It:

Vinegar (*acetic acid*) reacts with the *calcium carbonate* and makes the eggshell disappear. The egg remains intact because of a thin membrane around the outside.

Teacher Tips:

Safety Tip: Be sure and wash hands after handling raw eggs. They may contain harmful bacteria.

0-7424-2746-3 *Hands-On Chemistry Experiments*

Name _____

Date _____

Eggs-periment Record Sheet

Draw how the egg looked.

At first	**After a few minutes**
After several hours	**The next day**

0-7424-2746-3 *Hands-On Chemistry Experiments*

Experiment 34 · Hidden Picture

Substances react with other substances—observing a chemical reaction.
How do oxygen and lemon juice (citric acid) interact?

What You Need:

- Fresh lemon
- Cotton swabs
- Cup
- Desk lamp or blow dryer
- Copy of page 79 for each student

What to Do:

1. Squeeze the juice from the lemon into a cup.
2. Use the cotton swab to draw a picture on page 79. Dip the swab into the lemon juice frequently.
3. Let the paper dry completely.
4. Turn on the lamp and hold the picture over the bulb for a few minutes or blow-dry the paper.

Let's Talk About It:

Can you see your picture? The heat helps the water to evaporate completely. Then the oxygen in the air reacts with the citric acid from the lemon juice and turns it brown.

Teacher Tips:

Safety Tip: An adult should heat the pictures.

Name _____

Date _____

Hidden Picture

Draw your picture here with the lemon juice.

0-7424-2746-3 *Hands-On Chemistry Experiments*

After making homemade glue, compare it to a variety of different glues such as rubber cement, glue sticks, etc.

After popping the top off the film canister, try the same activity using other liquids such as vinegar, lemon juice, milk, etc. See which reacts more quickly.

Try adding a teaspoon of salt to a carbonated beverage to see a similar reaction to that in *Fizz*. You can also try doing the *Fizz* activity with several glasses at once using different colors. Let the colored fizzes combine. Do the colors mix?

After testing the four liquids for reaction, test other liquids.

Demonstrate the use of the yeast reaction by making a simple yeast bread or roll recipe. Allow students time to observe the yeast reaction when you combine the yeast, sugar, and warm water.

After *Eggs-periment*, you can try soaking a hard-boiled egg overnight. It will turn rubbery.

Use lemon juice to write secret messages. Does lime juice or orange juice work as well? What about milk?

                                  0-7424-2746-3 *Hands-On Chemistry Experiments*